by Mark Fara

Copyright © by Harcourt, Inc.

All rights reserved. No part of this publication may be reproduced or transmitted in any form or by any means, electronic or mechanical, including photocopy, recording, or any information storage and retrieval system, without permission in writing from the publisher.

Requests for permission to make copies of any part of the work should be addressed to School Permissions and Copyrights, Harcourt, Inc., 6277 Sea Harbor Drive, Orlando, Florida 32887-6777. Fax: 407-345-2418.

HARCOURT and the Harcourt Logo are trademarks of Harcourt, Inc., registered in the United States of America and/or other jurisdictions.

Printed in Mexico

ISBN 978-0-15-362484-1
ISBN 0-15-362484-1

2 3 4 5 6 7 8 9 10   126   16 15 14 13 12 11 10 09 08

Visit *The Learning Site!*
www.harcourtschool.com

Introduction

What do you think of when you think of robots? Maybe you think of a scene like this, from a play written in 1920 called *R.U.R.*: A weird scientist invents a strange material and uses it to make artificial people he calls "robots." Countries buy armies of these robots and use them as soldiers. The robots take over the robot factory and make more robots. At the end of the play, only robots are left.

This scary scene can't really happen—at least not today. About 50 years ago, robots were added to the list of real things that used to be science fiction. When you think of robots, you probably picture walking, talking, mechanical people, the kind you sometimes see in movies and on TV. That kind of robot has not been invented—yet. But robots have improved a great deal since the 1950s and can do many remarkable things.

Are There Robots in Your Home?

Do you ever use a microwave oven? When you watch TV, do you ever use a remote control to change channels or adjust the volume? Do you have

Robots from *R.U.R.*

a VCR that you can set to record a show? If you answered "yes" to any of these questions, you have robots in your house!

A robot is a machine that is programmed to do work that humans do. They also do jobs that humans cannot do. When something is programmed, it is given instructions to perform a task. Some robots can do very complicated tasks, but they can do only the tasks they are programmed to do.

When you put a hot dog in the microwave and set the timer, you are programming the oven to cook it and to let you know when it is done. Your TV remote is programmed to set your TV by following your instructions. Your VCR or DVR records your show while you are busy doing something else because you program it ahead of time. These are all examples of robots doing jobs according to your instructions. Other robots do work that is boring or dangerous for humans to do. These robots can assemble cars, help doctors with medical procedures, explore the sea and outer space, and perform many other tasks.

Where Did the Idea for Robots Come From?

Although the word *robot* is less than 100 years old, the idea of artificial people is much older. The ancient Greeks had myths about a god called Hephaestus, who built mechanical people to help him. Others throughout history have created myths, legends, and stories about humanlike beings that were built instead of born.

In the 1600s and 1700s, some people made *automatons.* Automatons were machines made in the shape of people or animals. The machines were made of gears, springs, and wheels, like an old-fashioned watch. When automatons were wound, they went through motions that copied the actions of real people and animals.

Some people thought automatons were creepy. The idea of a machine that could act like a person was scary! There is a story about a mathematician who created an automaton in the shape of a little girl. It scared some sailors so badly that they threw it off a boat and into the ocean. Another man who built an automaton that could write sentences was thrown in jail, along with his machine. People thought that he had used magic to make the automaton move.

Some old automatons still survive. They can be fascinating, but they are not robots. They are only wind-up dolls. They entertain people, but they do not do anything useful.

When Were Real Robots First Invented?

Like many inventions, robots developed over a period of time and involved ideas from many people. More than 100 years ago, a man named Nikola Tesla was the first to design a machine that could be operated by remote control. Tesla believed that it would someday be possible to make machines that were as smart and as useful as people.

From the late 1920s until the 1940s, a company called Westinghouse experimented with building mechanical people. The results were machines that were somewhere between automatons and robots. The most interesting machines were a mechanical person named Elektro and a metal dog, Sparko. Elektro walked and talked in preprogrammed ways when its inventor gave orders. Sparko followed Elektro, wagging its tail as it barked and growled. Like automatons, they were entertaining but of no real use.

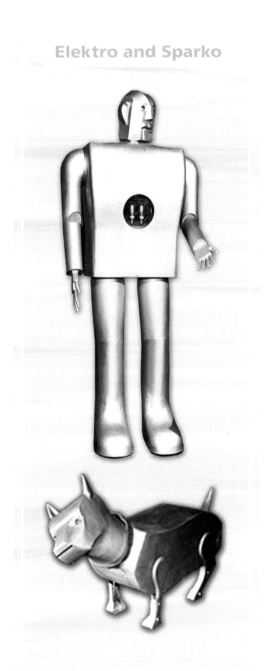

Elektro and Sparko

The first invention that was truly a robot was built in 1954. An inventor named George Devol made a machine called a manipulator. It looked and acted like a human arm, and it could be programmed to do many jobs. Devol helped start the first robot company in 1956. By the early 1960s, robots like the manipulator were being manufactured and sold to factories. This same kind of robot, with some improvements, is still widely used today.

In the late 1960s, a robot on wheels named Shakey was invented. Shakey was the first robot that was considered intelligent. Its *brain* was a huge computer that communicated with it by radio. At about the same time, an engineer built a 3,000-pound, four-legged robot that could walk up to 4 miles an hour. It was called the Walking Truck. Shakey and the Walking Truck could understand a few simple commands and move, but they were not intelligent in the same way that people are.

Robots Today

Right now there are more than 1,000,000 robots being used in the world. There are about 175,000 robots in the United States alone. Europe uses about twice as many as America does. Japan uses the most robots—more than half a million! Most robots today are run by electric motors and controlled by computers. Like earlier robots, these robots cannot think for themselves. They can only follow programs. But people have developed programs and robots that can do many things.

Think of a task. If there is not a robot that can do it, there is probably somebody trying to invent one. As scientists and engineers develop better robots, we will be seeing them doing more and more. In the future, it will not be unusual to see robots making highways, constructing buildings, laying underground pipe, and performing many other difficult tasks.

Since 1975 there have been robotic arms that can pick up objects, move them, and set them down exactly the way the programmer wants. Many of today's robotic arms can telescope. This means that they can become longer or shorter when needed. Some robotic arms can bend like the hose of a vacuum cleaner. Many of them have "hands" that work the way human hands do. These hands are called *grippers.* Different types of grippers are made, and each type has a specific purpose. Some of them can screw on bolts. Others can carry test tubes or hold welding rods.

Factory robots at work

Factory Robots

Factory robots are used on assembly lines, where the same parts need to be put together for each thing being made. Factories also use robots to paint, weld, and move parts from one place to another. One large car maker uses more than 16,000 robots in its factories. Today, almost every car made in the world is made with some robotic help.

Safety is a major reason for robot use in factories. Doing the same thing again and again for hours and hours is very dull. When people do this, they become bored and tired. After a while, it is hard for them to pay attention. People start making mistakes. Sometimes they injure themselves. None of this is a problem for robots.

Robots do a job exactly the same way every time. People may change how they do the job—even without realizing it. Robots do exactly what they are programmed to do, so they go through exactly the same actions every time. They do the work more consistently than people do.

In addition to being safer and more consistent, factory robots have another advantage—they are not living things. They never get sick or need lunch or restroom breaks. They do not need vacations or paychecks. They can work 24 hours a day without sleeping. Keeping a robot in good working order is cheaper than hiring people to do the same job.

This can be both good and bad. Using robots makes it cheaper and faster to produce goods. Companies can sell their products for less money, and more people can afford them. But people who are replaced by robots have to find new jobs. Fortunately, some new jobs are created by the use of robots too—jobs such as programming robots. These jobs are usually more complicated than factory work. People often need special training to do this kind of work.

Robots in Dangerous Places

Robots can go places that humans cannot go, such as deep into the oceans, inside volcanoes, and into outer space. Submersibles are robots that can be sent under water. They can locate sunken ships or look for valuable minerals. The most famous underwater robot is named *Jason.* It is owned by a group of underwater explorers in Massachusetts. *Jason* is the size of a small car and uses lights and cameras to transmit information back to the surface. Jason discovered the *Lusitania,* a famous ship sunk by a German torpedo in 1915. It has also found rare and valuable objects in ancient ship wrecks. Other submersible robots have explored sea life at depths where human divers cannot go safely.

Robots like *Dante II* (shown on the right) go into volcanoes to gather information and to help predict future eruptions. *Dante II* had legs that allowed it to move over the steep slope of a volcanic crater. It used cameras to keep track of where it was, and to send data back to scientists. Scientists used computers and satellites to control and communicate with the robot. NASA used *Dante II* to test ideas for robots that can be used on other planets.

Dante II was a volcano-exploring robot.

Since the 1960s, robots have been used to explore and learn about outer space. Sending robots into space is less expensive than sending astronauts, and human life is not at risk. In 1996, a space probe called *Galileo* went to Jupiter. Robots on this probe gathered information about Jupiter's atmosphere and sent the data back to Earth. Robots have also been used to explore and look for life on Mars and on one of Saturn's moons. Robots can take samples of soil and send information about the samples back to scientists on Earth. They also photograph other planets and take many measurements. Scientists send commands by radio waves that control computers on the spacecraft. The computers steer the craft and move its scientific instruments, as well as record data to send back to Earth.

Robots are also useful for going into areas on Earth that are dangerous for people. They can be used to clean up toxic chemicals and spilled nuclear waste. Some police departments use robots to defuse bombs. These robots have very flexible "fingers" and are run by remote control. If a bomb explodes, people are not killed or injured.

Robots can be made very small, allowing them to be sent into spaces that are too small for people. Small robots with cameras are often used to search for people in collapsed buildings. Survivors can be rescued more quickly because the robots can show rescuers exactly where to look.

Military Robots

Government scientists are developing robots that can find their way around without being programmed by people. These machines would be used in combat to carry food and supplies to soldiers, and to carry wounded soldiers to safety.

The Army already has a plane called the *Predator,* which is run from the ground by remote control. It is used to fly over dangerous places and take pictures. It can give the Army valuable information without risking the life of a pilot. It can be used to find people and to make maps that soldiers also can use.

Researchers are also working on robots that can be used to guard military bases and other important buildings. These robots would be able to patrol and explore dangerous areas and to follow suspicious people. They would be able to climb stairs, open doors, and locate booby traps. Using robots would protect people from being injured. Robots may also be used to clear away land mines and underwater mines.

Medical Robots

In the 1800s, a famous writer named Edgar Allan Poe wrote one of the first stories that hinted at a practical use for robots. It was called "The Man That Was Used Up." It was about an injured person who had his arms and legs replaced by machines. This was considered strange and unlikely in Poe's time, but it has turned out to be another example of science fiction becoming scientific fact.

In 1963, the Rancho Arm was invented to aid people with disabilities. It was an artificial arm controlled by a computer. It had six joints, making it as flexible as a human arm. Since then, artificial limbs, called *prosthetics*, have been developed that are even more useful. Today people who have lost arms, legs, or hands to injuries or disease can move and do many things with the aid of advanced prosthetics.

Robots also help medical workers in many ways. Some are used in hospital laboratories. Because robots cannot catch diseases, they are used to handle blood and other biological material that can spread germs. Robots are also sometimes used by surgeons. The steady movement of a robot can be very useful when a patient needs delicate surgery. Eye surgery is an area where robots are especially useful because it includes small movements that must be precise.

Some researchers are working on robots that can perform surgery while being controlled by a doctor hundreds or thousands of miles away. Other researchers are making progress in creating very small robots. Someday soon, there may be robots that are so tiny that they can barely be seen. These robots would be injected into people, just the way you get a shot to prevent illness. Once inside a person, they would be able to deliver medicine or clean out arteries to help prevent heart attacks and strokes.

Household Robots

You have read about some of the robots used in people's homes. More are on the way. There are inexpensive robots that can play soccer on your coffee table or kitchen floor. Some people have toy robotic dogs that can bark and follow them. You may even have built your own robots using construction toys! These robots respond to commands that people give. Before long, you may have robots in your home that can run on their own after you start them.

There is no such thing as a robot maid or butler—yet. You cannot find a robot that will walk with you to school. There is no place to buy a robot that will put away the dishes or help you do your homework. But things are starting to move in that direction. A company in Massachusetts has already invented a small household robot that will vacuum your floor!

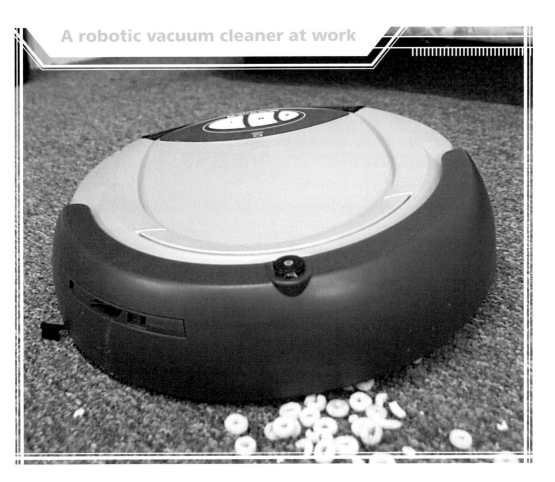
A robotic vacuum cleaner at work

The robot costs about the same as a regular vacuum cleaner. When it is turned on, it moves around the room and vacuums. It is programmed to keep track of where it has been, and when it has covered the entire floor a few times it beeps. Then it turns itself off. It can tell when it is near a wall or another barrier, so it won't crash into things.

Sometimes the vacuum robot leaves a little dirt in the corners. Once in a while it gets stuck under the sofa. And it is a good idea to keep the doors shut so that it does not wander off. But even though the robot is not perfect, it may be a preview of household robots to come.

How Can a Robot Know What Is Going on Around It?

The vacuum robot can sense when a wall is near. Robots used to explore space can keep themselves pointed in the right direction. To do all these things, the robot has to be aware of its environment. Human beings have senses. We can see, hear, smell, taste, and touch. A robot needs to be able to get the same kinds of information.

This problem has already been addressed. Robots can be equipped with sensors, circuits, and a microcontroller. Sensors are electronic devices that identify things the way human senses do. For example, a robot can "see" using optical sensors. An optical sensor detects the amount and type of light. It also detects shapes and colors. It converts what it sees into electrical impulses.

These impulses travel inside the robot on a path called an electric circuit. The circuit is a pathway through which electricity can travel. The circuit carries the impulses to a microcontroller. A microcontroller is a small computer that understands what the impulses mean. It uses the information and compares it to the robot's programming. It then sends electrical signals over the circuits back to the robot's sensors or other parts. The robot responds according to the instructions from the microcontroller. You might say that a microcontroller does the same thing for a robot that a brain does for a person. The circuits act like the nerves in the human body.

Sensors can also work like the human sense of touch. Small switches close when pressure is put on them. When they close, they send impulses through circuits to the microcontroller in the same way that optical sensors do. Another type of sensor can serve as "ears" for a robot. Sound impulses are carried to the microcontroller, which recognizes them and follows programmed instructions when it "hears" the sounds. Sensors such as this can also be used to determine how far away an object is. The robot emits a sound and "listens" for its echo bouncing off an object. It uses the time it takes the echo to bounce back to identify how far away the object is. This process is like sonar. It is the same method bats use to fly in the dark without crashing into things.

Movement is another challenge for robots. It is very hard to make a robot that can walk on two legs without falling over. Walking is really a complicated process involving balance and the use of many muscles. People do it without thinking. Robots can't.

Programming a robot to walk isn't easy.

There has been some progress in creating walking robots. A company in Japan has invented robot puppies. Another has designed small, human-looking robots that can jump, dance, and kick a soccer ball. One American company has made a life-size, human-shaped robot that can walk well as long as the surface it is walking on is smooth. But when there is a hill, the robot falls over. It also has a very hard time turning corners. Only recently was a company in Japan able to build a robot that can walk and run. Much more work needs to be done on this problem.

Are There Any Robots That Can Think?

Right now, even with sensors, a robot can only do its tasks according to instructions it has been given. It can't come up with its own response to anything. It can only follow its programming. No one has figured out how to make a robot think or learn.

Many researchers are working hard to develop machines that can think. Their goal is to invent artificial intelligence, or AI. AI would be to robots what intelligence is to people. It would let robots think, act, and learn on their own. Scientists have been working on AI since the late 1940s. With the invention of computers, the creation of AI seems much closer.

Scientists hypothesize that there are two steps to creating AI. The first is to make a database that contains all of the knowledge that is in a human brain, arranged in such a way that a machine can use it. But there is so much knowledge in a human brain that doing this might be impossible.

Some scientists are working at building databases with a lot of information about a single subject. But even this is a huge task. If the subject is detailed, like biology, there is an enormous amount to know. A subject like chess is easier because there is much less to know. A person playing chess with a robot would first make a move. The robot would then scan a database with a list of every possible move and pick the best counter move.

So far, so good—but then it gets trickier. How can a machine learn? In our example, the robot would not only have to choose its moves, it would need to realize when it made a bad move and know not to do it again. By learning from its mistakes, it would eventually know so much about playing chess that it would never lose a game.

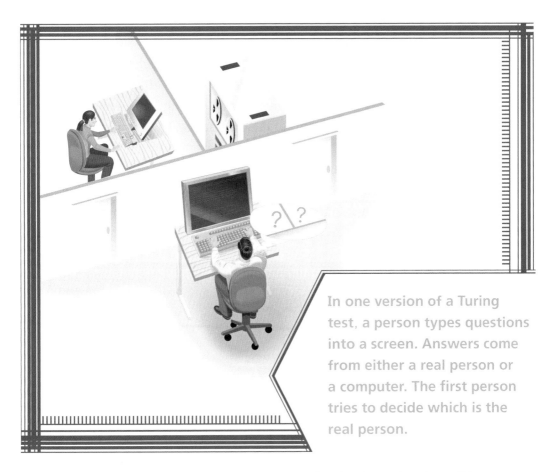

In one version of a Turing test, a person types questions into a screen. Answers come from either a real person or a computer. The first person tries to decide which is the real person.

But this robot would only be able to use its database for chess. If someone put a table tennis ball on the chessboard, the robot's microcontroller would not know what it was or what to do about it. Information about table tennis balls would have to be added to its database. If someone stuck a piece of tape on the chessboard, the robot would have to be programmed to recognize tape—and so on. Designing machines that can learn is one of the great challenges facing scientists who study AI.

Scientists working on AI are also unsure how they would know whether a robot was really thinking. In 1950, a scientist named Alan Turing tried to answer this question. He came up with a simple test. If a machine can perform an intelligent act in the same way that a person can, he said, then it is thinking. Many scientists agree with Turing, but some do not. Some believe that even if a machine acts as if it is thinking, it does not understand what it is doing. This question may not be settled for a very long time.

Conclusion

All over the world, inventors are working hard to develop truly intelligent, humanlike robots. There are already robots all over the place, small ones in our homes and larger ones that do boring, dangerous work for us. We have robots that can walk, talk, and dance, but they cannot really think—yet.

For now at least, we do not need to worry about robots ruling the world. They would need to be able to think and solve problems the way a person can, and none of them can do that. But we should not take robots lightly. They make life easier for us in many ways.

Is there one of these in your future?